Strategic Studies Institute
and
U.S. Army War College Press

SPACE-BASED SOLAR POWER: A TECHNICAL, ECONOMIC, AND OPERATIONAL ASSESSMENT

Jeffrey L. Caton

April 2015

*****

Comments pertaining to this report are invited and should be forwarded to: Director, Strategic Studies Institute and U.S. Army War College Press, U.S. Army War College, 47 Ashburn Drive, Carlisle, PA 17013-5010.

*****

This manuscript was funded by the U.S. Army War College External Research Associates Program. Information on this program is available on our website, *www.StrategicStudiesInstitute.army.mil,* at the Opportunities tab.

*****

*****

The Strategic Studies Institute and U.S. Army War College Press publishes a monthly email newsletter to update the national security community on the research of our analysts, recent and forthcoming publications, and upcoming conferences sponsored by the Institute. Each newsletter also provides a strategic commentary by one of our research analysts. If you are interested in receiving this newsletter, please subscribe on the SSI website at *www.StrategicStudiesInstitute.army.mil/newsletter.*

ISBN 1-58487-668-9

FOREWORD

With growing international awareness of energy security challenges, the promise of space-based solar power for clean and unlimited energy for all humankind is certainly appealing. While significant progress continues in the enabling technologies of such systems, is there compelling evidence that space-based solar power systems will provide the best energy solution? How does the Army's current approach to incorporating a diverse portfolio of renewable energy sources in distributed locations compare with the potential of enterprise ventures that beam energy from solar collectors in space?

For more than 4 decades, many credible organizations in government and industry have explored the concept of space-based solar power. But their serious studies often conclude that such systems remain on the future horizon, usually at least 10 years away from practical application. This monograph posits that, while space-based solar power systems may be technically feasible, there is no compelling evidence that such systems will be economically or operationally competitive with terrestrial-power generation systems in use or in development. However, this monograph does find that there may be some utility in the limited application of space-based solar power to enable operations in remote and forward operating locations.

DOUGLAS C. LOVELACE, JR.
Director
Strategic Studies Institute and
U.S. Army War College Press

ABOUT THE AUTHOR

JEFFREY L. CATON is President of Kepler Strategies, LLC, Carlisle, PA, a veteran-owned small business specializing in national security, cyberspace theory, and aerospace technology. He is also an Intermittent Professor of Program Management with Defense Acquisition University. From 2007 to 2012, Mr. Caton served on the U.S. Army War College faculty, including Associate Professor of Cyberspace Operations and Defense Transformation Chair. Over the past 5 years, he has presented lectures on cyberspace and space issues related to international security in the United States, Sweden, the United Kingdom, Estonia, and Kazakhstan, supporting programs such as the Partnership for Peace Consortium and the North Atlantic Treaty Organization Cooperative Cyber Defence Center of Excellence. Mr. Caton's current work includes research on cyberspace and space issues as part of the External Research Associates Program of the Strategic Studies Institute as well as serving as a facilitator for Combined/Joint Land Force Component Commander courses at the Center for Strategic Leadership and Development. He served 28 years in the U.S. Air Force in engineering, space operations, joint operations, and foreign military sales, including command at the squadron and group level. Mr. Caton holds a bachelor's degree in chemical engineering from the University of Virginia, a master's degree in aeronautical engineering from the Air Force Institute of Technology, and a master's degree in strategic studies from the Air War College.

SUMMARY

The concept of generating electrical power from solar energy using satellites and then transmitting that power to Earth is decades old and generally considered to be technically feasible. If successful, such systems could provide constant access to almost unlimited power and thus play a significant role in U.S. national and international energy security strategies. However, the practical application of this method of power generation requires economical and operational feasibility as well. This monograph examines the current progress of space-based power in these three areas: technology, economy, and operations. The scope of discussion is at the survey level of detail to provide senior policymakers, decisionmakers, military leaders, and their respective staffs an overall appreciation for the challenges, opportunities, and risks associated with space-based solar power systems.

This monograph has three main sections:

1. Technical Assessment. This section introduces the basic concept of space-based solar power (SBSP). It then summarizes the evolution of the concept's development, as documented in six major reports written over the past 37 years. Finally, the section examines the critical technologies required for the successful development of the space, ground, and support elements of the system.

2. Economic Assessment. This section examines SBSP system cost estimates from a variety of sources. It then compares these costs to competing alternative energy solutions such as terrestrial-based photovoltaic power plants. The section also addresses regulatory factors that may affect the development and operation of SBSP systems as well as current international efforts in this field.

3. Operational Assessment. This section explores the strategic considerations for SBSP systems within the general context of national space operations. It then examines potential garrison-level applications and compares these with the current plans of the Army's Energy Initiatives Task Forces to integrate terrestrial-based photovoltaic power into the energy systems of several major installations. Finally, the section briefly explores possible SBSP applications to support remote operating locations.

SPACE-BASED SOLAR POWER: A TECHNICAL, ECONOMIC, AND OPERATIONAL ASSESSMENT

The concept of generating electrical power from solar energy using satellites and then transmitting that power to Earth is decades old and generally considered to be technically feasible. However, the practical application of this method of power generation requires economical and operational feasibility as well. This monograph examines the current progress of space-based power in these three areas: technology, economy, and operations.

TECHNOLOGICAL ASSESSMENT

This section introduces the basic concept of space-based solar power (SBSP). It then summarizes the evolution of the concept's development, as documented in six major reports written over the past 37 years. Finally, this section examines the critical technologies required for the successful development of the space, ground, and support elements of the system.

Concept.

The idea of harnessing the power of the sun in space and transferring that energy to Earth is older than the first U.S. satellite. Initially conceived in the realm of the imagination, the idea soon gained the attention of serious scientists. In 1968, Dr. Peter Glaser publicly presented his research in the area of space-based solar power at the Intersociety Engineering Energy Conversion Conference.[1] Three years later, he filed for a U.S. patent for a "Method and Apparatus for Converting Solar Radiation to Electrical Power," which was granted in 1973. In the patent document, Glaser emphasized the problems of pollution and waste management caused in energy production using fossil and nuclear fuels as well as the limited supply of each on the planet. He argued that his invention would overcome many of the practical limitations of generating electricity from solar power on the Earth, such as atmospheric absorption, cloud cover, and day-night cycles.[2]

The abstract for Glaser's patent provides a concise description of the process and its key functional elements: "Solar radiation is collected and converted to microwave energy by means maintained in outer space on a satellite system. The microwave energy is then transmitted to Earth and converted to electrical power for distribution."[3] Figure 1 provides a simple block diagram of this design with elements grouped by space systems (collect, convert, and transmit) and terrestrial systems (receive, convert, and utilize). These systems are connected by an electromagnetic (EM) link to achieve energy transfer using microwaves per the original design or using light amplification by stimulated emission of radiation (laser) as proposed in certain subsequent designs. The technological assessment in this paper will focus on these seven functions.

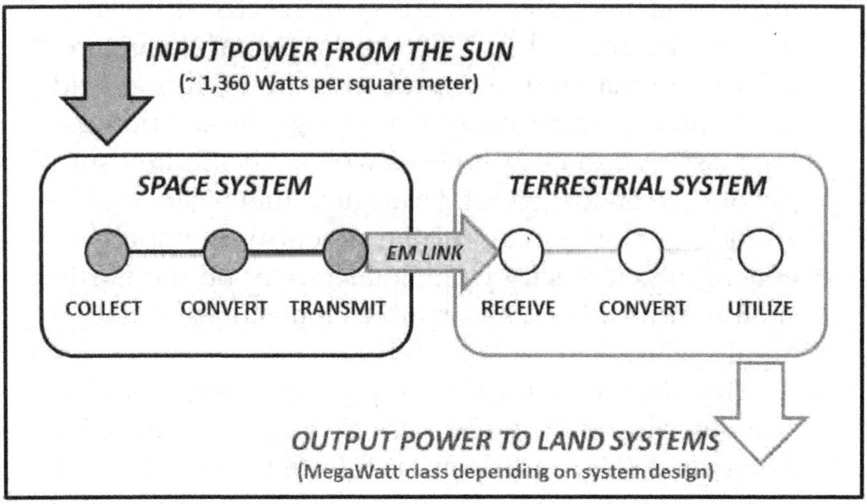

Figure 1. Key Functional Elements of a Space-Based Solar Power System.[4]

Evolution and Trends.

The U.S. Government's interest in SBSP systems has waxed and waned over the years in large part due to the lack of urgent need for the capability—coupled with very large investment requirements competing with other national needs. Hundreds of studies and reports have been written on the various details of the SBSP concept, but most of the content relevant to this monograph is contained in six comprehensive studies summarized here.

Initial National Aeronautics and Space Administration Reports (1977 and 1980).

Less than 2 years after Glaser's initial patent was granted, the National Aeronautics and Space Administration (NASA) George C. Marshall Space Flight

Center initiated a contract in February 1975 for a three-phase study of SBSP concepts. The final report, "Space-Based Solar Power Conversion and Delivery Systems Study," was published in March 1977. The study "examined potential concepts for a photovoltaic satellite solar power system, focusing on ground output power levels of 5,000 megawatts (MW) and 10,000-MW, and a power relay satellite," as well as economic issues related to their implementation.[5] If one considers that the 2012 net generation capability of the Three Mile Island nuclear plant was 829-MW, the scope of application envisioned by the NASA report was significant.[6]

The study had three major conclusions: (1) SBSP "is technically feasible and has economic potential"; (2) the economic potential provides justification for "a significant technology advancement and verification program"; and, (3) there remained "major areas of technological and economic uncertainty relating to decisionmaking." The report highlighted four areas requiring further study:

> a) The fabrication and assembly of large structures in space; b) Solar energy conversion technology; c) The cost of electric power supplied by alternative energy sources; and, d) Constraints imposed by ionospheric and biological effects.[7]

As we will see, these specific areas of challenge remain largely unresolved by any practical demonstration. The cost estimate to procure a 5,000-MW SBSP system and deliver it to geosynchronous orbit (GEO) was $7.566 billion, with an estimated annual operations and maintenance cost of $1.156 billion (both costs based on 1974 dollar value). Of particular note for further discussion is that the costs for the space launch of the system ($3.278 billion) comprised over 43 percent of the total procurement cost.[8] The study also anticipated a constellation of 109 to 120 satellites, each with a service life of 30 years, with the prototype satellite on orbit within 10-15 years (1987-92).[9] Overall, this study was extremely ambitious in its operational scope as well as very optimistic regarding the potential availability of funding.

In the same year as the release of the first NASA SBSP report, the Department of Energy (DoE) initiated their Concept Development and Evaluation Program (CDEP) to work jointly with NASA in further examination of solar power satellite costs and benefits. In 1980, DoE published its final report of the Solar Power Satellite (SPS) Program Review. Open participation was encouraged from a broad audience, and the report included an impressive collection of over 170 separate articles by subject matter experts organized into four general areas: Systems Definition, Environmental, Societal, and Comparative Assessments. To help integrate the myriad technical details of the individual studies, the report established a baseline SPS configuration that many subsequent reports refer to as the "1979 SPS Reference System." This system's major elements (depicted in Figure 2) are truly monumental in scale, with each satellite's solar collector measuring 50 square kilometers (KM), a transmitting space antenna that is 1-KM in diameter, and a terrestrial receiving antenna array measuring 10 by 13-KM.[10] Despite having an estimated price tag of $13.5 billion (1977 dollar value) for the first satellite and an eventual unit cost of $11.5 billion—plus an estimated $89 billion in development costs—the report "bottom line" was that "no insurmountable barrier to the SPS concept have [sic] emerged. Significant technical, environmental and cost questions, however, have not been answered."[11]

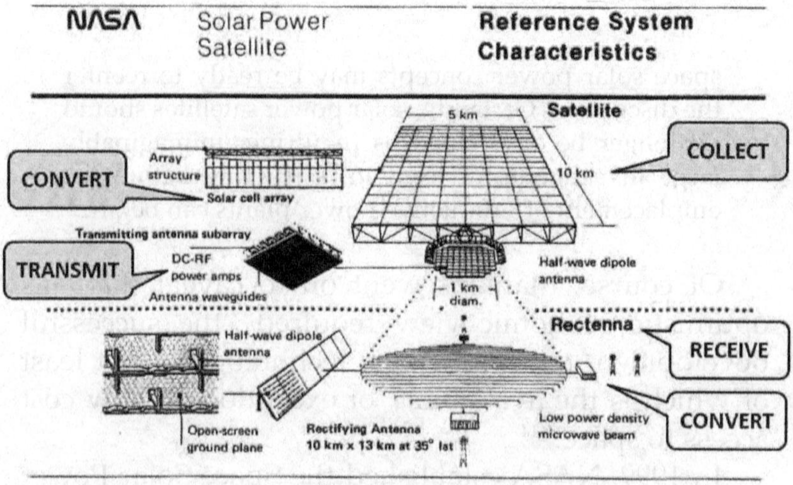

Figure 2. 1979 NASA/DoE Solar Power Satellite Reference System.[12]

NASA "Fresh Look" Report (1997).

After the Congressional Office of Technology Assessment determined in 1981 that the DoE/NASA SPS concept was "programmatically and economically unachievable," U.S. Government efforts for SBSP development went dormant. During 1995-96, NASA decided to examine the issue again from a "fresh look" perspective to see if new technologies might provide new solutions to existing SBSP challenges. The study considered 30 different SPS models as well as different orbital architectures, which included low Earth orbit (LEO) and medium Earth orbit (MEO) in addition to the 1979 standard placement in GEO. It re-addressed the major parts of the original DoE/NASA study including the space system, ground system, space infrastructure, and space transportation. Renowned SBSP expert and advocate John Mankins summed up the study's conclusion:

> space solar power concepts may be ready to reenter the discussion. Certainly, solar power satellites should no longer be envisioned as requiring unimaginably large investments in fixed infrastructure before the emplacement of productive power plants can begin.[13]

Of course, Mankins went on to caveat that this optimistic economic view required "the successful development of various new technologies—not least of which is the availability of exceptionally low cost access to space."[14]

In 1999, NASA established the Space Solar Power (SSP) Exploratory Research and Technology (SERT) program to further explore some of the claims of the "Fresh Look" report, focusing on technologies needed to make the concept competitive with traditional ground-based power plants. In addition, it was:

> also expected to provide a roadmap of research and technology investment to enhance other space, military, and commercial applications such as satellites operating with improved power supplies, free-flying technology platforms, space propulsion technology, and techniques for planetary surface exploration.[15]

Accordingly, SERT used a series of model system categories (MSCs) of experimentation demonstrations to integrate its research efforts. The MSC designs ranged from relatively small-scale satellites to deliver 100-KW of power from LEO; to megawatt-range satellites that could "fly" from Earth LEO to GEO as well as interplanetary; up to the gigawatt (GW)-range satellites required for commercial power generation. The overall goal was to use a constellation of 1.2-GW satellites to deliver 10-100-GW of power for commercial use on Earth. Even with an extremely optimistic space lift cost, each satellite was estimated to incur $14 billion for transportation to operational orbit.[16] In March 2000, NASA asked the National Research Council to conduct an independent assessment of its SBSP programs; the group's final report concluded:

> The committee has examined the SERT program's technical investment strategy and finds that while the technical and economic challenges of providing space solar power for commercially competitive terrestrial electric power will require breakthrough advances in a number of technologies, the SERT program has provided a credible plan for making progress toward this goal. The committee makes a number of suggestions to improve the plan, which encompass three main themes: (1) improving technical management processes, (2) sharpening the technology development focus, and (3) capitalizing on other work. Even if the ultimate goal—to supply cost-competitive terrestrial electric power—is not attained, the technology investments proposed will have many collateral benefits for nearer term, less-cost-sensitive space applications and for nonspace use of technology advances.[17]

National Space Security Office Report (2007).

In March 2007, the National Space Security Office (NSSO) Advanced Concepts Office initiated a collaborative study to examine the potential benefits of SBSP systems from the perspective of national energy and enviromental security threats and opportunities. Its approach "relied heavily upon voluntary Internet discussions by more than 170 academic, scientific, technical, legal, and business experts around the world."[18] The stated research question was:

> Can the United States and partners enable the development and deployment of a space-based solar power system within the first half of the 21st century such that, if constructed, it could provide affordable, clean, safe, reliable, sustainable, and expandable energy for its consumers?[19]

The report "Space-Based Solar Power as an Opportunity for Strategic Security" addressed many aspects of this question by building upon the conceptual frameworks established in early NASA and DoE reports. It also addresses the study's operational goal to deliver 5-10-MW of power to remote locations (vice designated terrestrial power stations) as well as the strategic goal to deliver 10 percent of all U.S. power by 2050.[20]

The NSSO report presented a more aggressive development option for a large-scale SBSP system demonstrator than the 25-year roadmap offered by NASA. NSSO's new roadmap consisted of three phases over 10 years, culminating in an SBSP satellite in orbit capable of delivering 10-MW of energy to Earth with a program cost of no more than $10 billion.[21] To provide a sense of scale for NASA's proposed system, the report noted that the largest structure currently in orbit, the International Space Station, had a mass of 232 metric tons (MT), which included solar panels that produce 112-KW of energy. The study's design was much larger; "a single Space Solar Power Satellite is expected to be above 3,000-MT, several kilometers across, and most likely be located in GEO . . . likely delivering between 1 to 10-GWe [gigawatts electrical]."[22] With such large and heavy satellites, it is reasonable the study considered "that launch cost is the single most important driver of the business case for SBSP," requiring these costs to be as low as $200 per pound to make the system competitive with traditional commercial energy provider rates of about 10 cents per kilowatt-hour.[23] NASA estimates the current price of space launch to be approximately $10,000 per pound to orbit, but its Advance Space Transportation Program aims to reduce this cost to hundreds of dollars per pound within 25 years.[24]

The report concluded that "the technical feasibility of the concept has never been better" but also admitted that "several major challenges will need to be overcome to make SBSP a reality, including the creation of low-cost space access and a supporting infrastructure system on Earth and in space." With regard to military operations, the report noted that "for the DoD [Department of Defense] specifically, beamed energy from space in quantities greater than 5-MWe has the potential to be a disruptive game changer on the battlefield." The report envisioned applications of SBSP energy being delivered to provide support across the spectrum of operations, such as providing energy on demand to forward-deployed combat units and ultra-long-duration surveillance drones, as well as rapid power delivery for disaster relief or nation-building efforts.[25]

Naval Research Laboratory Report (2009).

The 2007 NSSO report helped to motivate management at the Naval Research Laboratory (NRL) to examine the potential benefits of SBSP with a formal study "to determine if the NRL can offer a unique, cost-effective, and efficient approach to supplying significant power on demand for Navy, Marine Corps, or other DoD applications."[26] The final report findings were presented in four areas, three of which were similar to previous reports—concept feasibility, relevant research areas, and recommended course forward. In general, the report agreed that SBSP was technically feasible, but that there remained significant areas of technical, operational, and economic risk.[27] It also highlighted how NRL might help reduce some of these risks, noting that "research applied to the areas in which NRL has core competencies would yield substantial technological dividends for SBSP as well as other space and terrestrial applications."[28]

The NRL report included an area of findings dedicated to military operations scenarios. This included providing power to forward-operating locations that the report considered the best SBSP defense application; it was rated as possible for technical and economic feasibility, with development costs estimated at $10 billion and a development time of at least 5 years. An application with similar assessments was that of providing power to an unmanned air vehicle to prolong its dwell time. Providing power to individual warfighters was also considered, but it was assessed as unlikely for both technical and economic feasibility, due to power inefficiency, precise beam control requirements, and safety concerns for extended exposure to the energy transfer beam.[29] The full summary of possible military scenarios and their assessments is included as Appendix 1.

International Academy of Astronautics Report (2011).

In September 2007, the International Academy of Astronautics (IAA) completed a proposal for the first truly comprehensive international study of SBSP concepts. The goal of the study was:

> to determine what role solar energy from space might play in meeting the rapidly growing need for abun-

dant and sustainable energy during the coming decades, to assess the technological readiness and risks associated with the SPSS [space-based solar power systems] concept, and (if appropriate) to frame a national international roadmap.[30]

Two years later, the results of the study were presented in the International Symposium on Solar Energy from Space in Toronto, Canada. The study was comprehensive; its results were organized into seven areas: satellite system concepts; supporting systems; technology readiness and risk assessment; policy, legal, and regulatory considerations; market assessment and economics; preliminary system analysis; and international roadmap. However, the overall report summary sounded familiar:

> As of 2010, the fundamental research to achieve technical feasibility for the SPS was already accomplished. Whether it requires 5-10 years, or 20-30 years to mature the technologies for economically viable SPS now depends more on (a) the development of appropriate platform systems concepts, and (b) the availability of adequate budgets.[31]

The study's detailed findings included themes that were familiar from previous reports, such as "low-cost Earth-to-orbit transportation is an enabling capability to the economic viability of space solar power for commercial base load power markets."[32] The findings also addressed the desire for SBSP energy to integrate with existing terrestrial power distribution systems as well as concerns for the resolution of policy and regulatory issues surrounding the transmission of energy from space to the Earth. The proposed technology roadmap emphasized experiments and demonstrations and noted that:

> timely success seems more likely to result from cooperation in accomplishing R&D [research & development] objectives among governments, among industry players and among a broad range of government, corporate and academic organizations.[33]

NASA Innovative Advanced Concepts Program Report (2012).

Almost a decade after NASA's SERT program completed its 2-year study, the NASA Innovative Advanced Concepts (NIAC) Program approved a grant to explore a new SBSP design that might resolve any remaining technological and economic uncertainties. The concept that emerged was the Solar Power Satellite via Arbitrarily Large Phased Array (SPS-ALPHA), an innovative approach that leveraged modular design and autonomous robotics to address some of the challenges related to building large structures in space. It would use hundreds of thousands of standard hexagonal-shaped modules with various functions—reflecting, photovoltaic, or transmission—in concert with interconnecting devices and robotic arms to produce satellites of desired size and configuration. The envisioned assembly method would be cooperative behavior, such as that demonstrated by "a team of skydivers who cooperated to form quickly a large, complex structure during a jump."[34] The modular design could also reap the economic benefits of lower unit costs due to mass production and the small size of each module would simplify some of the physical challenges of space launch.

Potential applications proposed for SPS-ALPHA included not only commercial power generation, but also "National Security Premium Niche Power Markets" that:

> may emerge due to military operations, or because of a requirement for short-term emergency operations (e.g., to support relief operations in the aftermath of a major national disaster, such as an earthquake, a tsunami, etc.).[35]

In economic terms, the report's conclusion looks promising:

> The roadmap for SPS-ALPHA appears quite tractable programmatically: the hyper-modular architecture should enable fast-paced, relatively inexpensive steps forward, with a total cost for a scalable solar power satellite pilot plant of about $5B and the first full-scale SPS of roughly $20B. These numbers are substantial, but compare well to the reported $100B cost of the ISS or the earlier 1980s era estimates of roughly $1,000B to reach the first SPS [cost adjusted for inflation from c. 1980 to c. 2012].[36]

However, the report also admits that these estimates have a familiar assumption:

> As has been found in past studies and for other SPS concepts going back to the 1970s, ETO [Earth-to-orbit] transportation remains a critical factor in realizing economically viable SPS for terrestrial markets.[37]

Critical Technologies.

From the overview of previous major studies, it is clear that there is no single or best-design concept for SBSP systems. Results from the research and development of individual components and subsystems could fill libraries, but from this diverse work, certain tenets have emerged with regard to critical SBSP technologies.[38] This section addresses such matters for a typical system's key functional elements as depicted in Figure 1. It then introduces the major support systems required for practical SBSP operations and highlights the related unique or critical technologies. The focus is on the assessment of feasibility and operational appropriateness to support decisionmaking and not on the understanding of scientific or engineering nuances.

Space System.

From a survey of the literature, there is general agreement in the technical community that SBSP systems technically feasible and require no scientific breakthroughs to operate. This is not to say that all required technologies are readily available, but rather that there is high confidence that they can be developed for practical application in a reasonable time with reasonable risk.[39] Thus, the focus in space system design turns toward the identification and optimization of performance parameters to distinguish operational utility among many different design alternatives. For example, factors such as component durability and efficiency may be balanced against their cost and weight.

The components of a space system as depicted in Figure 1 involve relatively mature technologies and pose low risk for development and operation. Solar radiation collection will require large reflectors using established basic material and structural design practices, with risk increasing with the size of the assembly. Solar power conversion can leverage various proven forms of photovoltaics—individual cells that convert solar radiation to electricity—that have been used on satellites for over 50 years. The efficiency of this conversion is a driving factor in the selection, but these efficiencies will degrade with continuous exposure to the sun, with the performance reduction different for each cell design and material.[40] Unlike today's satellites, which use flat photovoltaic arrays almost exclusively, SBSP satellites would need to have the solar collectors concentrate sunlight to near the maximum capacity for the solar cells to further increase the overall system efficiency.[41]

The wireless energy transmission subsystem has received considerable attention in every major SBSP study. The two most popular choices are either microwave or laser transmission subsystems. Microwaves have the advantage of significantly better efficiencies but require very large transmitting antennae, which, in turn, drive the need for large receiving antennae on the ground. Lasers are more compact and require much smaller ground receivers, but this drives the need for more accurate beam pointing. Also, the lower efficiencies of laser operation mean that considerably more energy is wasted in the form of heat than with microwave systems, and this heat must be managed as part of the satellite operation.[42] One innovative solution to the transmission challenge is to combine it with the sunlight-conversion process in a modular design. The NRL has built and tested such a design called the "sandwich module," in which "one side receives solar energy with a photovoltaic panel, electronics in the middle convert that direct current to a radiofrequency, and the other side has an antenna to beam power away."[43] This type of module is an essential part of the SPS-ALPHA design configuration, which envisions the use of hundreds of thousands of such components for each satellite.[44]

While most of the SBSP components may be low risk individually from a technological viewpoint, the systems integration effort required to make a viable satellite that can operate for decades is immense. The satellite must have an integrated structure (commonly called the "bus") that includes power and thermal management systems; guidance, navigation, and control systems; telemetry, tracking, and control links; and stabilization and pointing systems. While it may be possible to leverage some of the design concepts of existing commercial GEO satellites buses, none of them can handle the size and mass of proposed SBSP applications.[45] The requirements of an SBSP satellite bus would be even more challenging for concepts based on lower orbits (e.g., LEO or MEO), in which orientation between the satellite and the ground stations is constantly changing, compared with the relatively stable orientation of GEO satellites.

Electromagnetic Link.

As mentioned, the electromagnetic (EM) link that provides wireless power transmission from the SBSP satellite and ground receiver would most likely be high-intensity microwave or laser beams. In theory, frequency ranges of 2.5 to 5.4 gigahertz (GHz) and 35 to 38-GHz are optimal for such application. While most studies favor the lower-frequency range, the higher range requires much smaller transmission and reception antennae and will produce less ionospheric heating.[46] Viable laser designs would operate in the near-infrared spectrum and would be generated using a solid-state electronic means vice chemical reaction. In either case, the operational design of the EM link must consider safety issues related to the propagation of such directed energy beams that may encounter other satellites, aircraft, or humans en route to its intended ground receiver.[47] The selection of the EM link source will have a significant impact on the design requirements of the satellite bus attitude control and stabilization systems, which are untried for such large structures. Even minor pointing and vibration errors become significant when they are projected across tens of thousands of miles.

Terrestrial System.

Compared to the space system, the terrestrial system generally has much lower technical risk. For microwave transmission, the receiver would be a rectified antenna, also known as a "rectenna," which would both receive the microwave energy and convert it into direct current electricity.[48] For laser transmission, the receiver would likely be a series of mirrors to concentrate the laser energy and direct it to photovoltaic cells for conversion to direct current electricity. In either case, the ground power management and distribution subsystems necessary to utilize the converted energy are common to the existing utility industry. Since there are inefficiencies inherent in either ground conversion process, the ground station must include thermal management to cool its components. Since the system is designed to always have access to energy from the sun, it may require temporary power storage on Earth for times of temporary satellite operation disruption or when the utility grid may not be able to accept all of the energy generated.[49]

Supporting Systems.

Thus far, we have only considered the elements of an SBSP system in a theoretical design configuration; to examine it in a realistic operational environment we must also consider the supporting infrastructure required to deploy, operate, and maintain the system. Figure 3 depicts the SBSP system block diagram within this more holistic context.

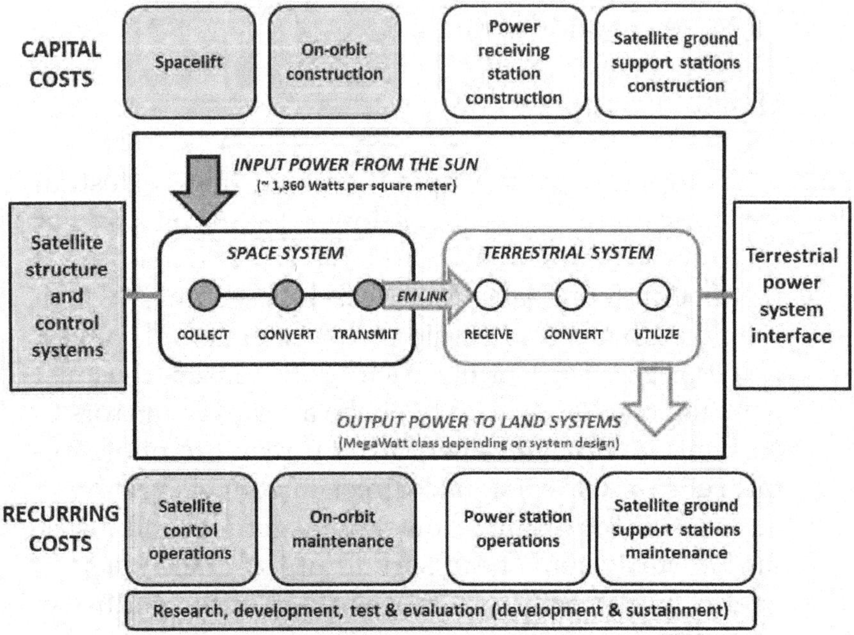

Figure 3. Space-Based Solar Power System with Supporting Infrastructure.

For the space system, the transportation to orbit of such a large satellite is unprecedented and may take literally hundreds of current state-of-the-art space launches to achieve.[50] The on-orbit assembly of the satellites is very high risk and deserving of its own detailed discussion beyond the scope of this monograph. The original NASA 1979 SPS Reference System included a reusable space freighter to carry astronauts to a

LEO construction depot where components were to be assembled and then transferred to GEO.[51] Contemporary studies have delegated these tasks to robotic devices yet to be designed, built, or tested. Once the system is fully deployed in space, it must be supported by routine satellite telemetry, tracking, and control (TT&C) facilities on Earth to monitor and maintain the health and station keeping of the satellite as well as direct operations of the SBSP payload. This is not a trivial matter; it requires ground facilities with TT&C transmitters and receivers as well as the equipment, procedures, and trained personnel to accomplish the tasks. Finally, the use of on-orbit maintenance would be necessary to ensure effective operation of the SBSP satellite with design lives of 10 years or more in the hazardous space environment that may induce damage through impact with micrometeoroids or masses of charged particles ejected by the sun.

The terrestrial systems infrastructure is more straightforward. It will require the construction and maintenance of a nature similar to that of existing photovoltaic solar power farms, but on a much greater scale. However, these systems may also require environmental impact studies of much greater detail than existing solar farms due to the increased intensity of the microwave or laser energy that they harvest. Of course, the terrestrials systems will also require the proper equipment, procedures, and trained personnel necessary for operations and maintenance. Finally, there needs to be a program of continuous research, development, test, and evaluation (RDT&E) supporting all of the space and terrestrial system activity. The RDT&E effort would not only support the initial design and deployment of the systems, but also would conduct programs such as component aging surveillance and technology insertion to optimize performance throughout the system life cycle.

ECONOMIC ASSESSMENT

This section examines SBSP system cost estimates from a variety of sources. It then compares these costs to competing alternative energy solutions such as terrestrial-based photovoltaic power plants. This section also addresses regulatory factors that may affect the development and operation of SBSP systems as well as current international efforts in this field.

Estimated Cost Assessments.

The cost estimates for SBSP systems included in some of the studies already discussed do not follow any standard format. Thus, a direct comparison of their values among alternate systems is not possible. Instead, let us assess the estimates that are available for their magnitude and completeness. Table 1 is a summary of the system-level cost estimates extracted from the indicated reports as well as from two recent master's degree theses. To aid with assessment, the original estimates have also been converted to 2013 dollars. In cases where a range of value was given for the estimate, the lower cost was selected; thus, the costs in Table 1 reflect the best-case scenarios. For comparison to contemporary utility rates, a cost for each kilowatt-hour (kWh) was calculated for both 10-year and 30-year system lives. To stay with the best-case theme, these values were calculated assuming continuous power product.

At first look, most of the designs have costs per kWh well under $1 over an expected duty life; some of them approach costs of 10 cents per kWh, which is currently a competitive rate for commercial residential power.[52] However, with the exception of the last estimate, all values in Table 1 do not include the costs to develop, build, and operate the terrestrial power plant—a significant cost for the overall system. Also, all the U.S. Government studies presented assume significant reductions (at least 50 percent) in space lift costs as part of their estimates. Thus, the SBSP systems do not yet achieve the economic feasibility necessary to compete with existing commercial power producers. However, the ability to deliver power to remote locations may open niche markets in which premium rates may be charged that allow for a profitable SBSP system.

Space-Based Solar Power (SBSP) System Description	Estimated Cost (year basis)	Estimated Cost (2013)[53]	Estimated Cost per kilowatt-hour (2013)	
			10-year life	30-year life
DoE/NASA 1979 SPS Reference System (5 GW)[54]	(1977)			
Satellite system (per unit cost)	11.5 B	35.0 B	0.08	0.03
Satellite development	89.0 B	270.6 B	0.62	0.21
Satellite annual operations	1.2 B	3.6 B	0.01	>0.01
NASA Fresh Look Study[55]	(1997)			
Sun Tower satellite (250-MW)	8 B	11.0 B	0.50	0.17
SolarDisc satellite (5-GW)	30 B	41.1 B	0.09	0.02
Launch cost to GEO[56]	14 B	19.2 B	0.18	0.06
NSSO Study[57]				
Development & demo satellite (10-MW)	(2007) 10.0 B	11.0 B	12.56	4.19
NASA NAIS SPS-ALPHA Study[58]	(2012)			
First full-scale satellite (1-GW)	20.0 B	20.0 B	0.23	0.08
Development roadmap	30.0 B	30.0 B	0.34	0.11

Table 1. Comparison of Space-Based Solar Power System Cost Estimates.

Naval Postgraduate School Thesis (Chow)[59] Alternative A satellite (110-MW) Alternative B satellite (106-MW) Alternative C satellite (65-MW)	(2013) 10.4 B 2.4 B 2.1 B	10.4 B 2.4 B 2.1 B	1.08 0.26 0.37	0.36 0.09 0.12
Toulouse Business School Thesis (Xin et al.)[60] (full system cost) European Space Agency (ESA) development (5-GW)	(2009) 265	275.9	0.63	0.21

Table 1. Comparison of Space-Based Solar Power System Cost Estimates. (cont.)

Even with the potential for profitable niche markets, the magnitude of investment required is staggering. The total development cost based on European Space Agency (ESA) estimates of over $275 billion (see last entry in Table 1) is more than the market value of Walmart Stores ($247 billion) and well beyond the fiscal year 2015 budgets for NASA ($17.5 billion) and the Department of Energy ($27.9 billion).[61] Further, the net present value (NPV) calculated on several SBSP systems yielded negative values: -$72 billion for a 1-GW system and -59 billion for a 5-GW system. The negative NPV values mean that the systems will subtract that value from the investor over the lifetime of the project.[62]

It is not clear that reducing the scope of the SBSP power generation to match needs of niche markets—for example, from 1-5-GW to 10-100-MW—would make the NPV calculations much better for investors, since many of the costs for research and development of the system components are the same.[63] In any case, there will always be competitors in the renewable energy market—how do their values compare to those of SBSP systems?

Competing Alternative Energy Solutions.

The DoE Energy Information Administration (EIA) produces independent statistics and analyses that include capital and operating cost estimates for utility scale electricity generating plants. Making these reports available to the public can play an "important role in determining the mix of capacity additions that will serve future demand for electricity." The analysis includes consideration of the overall context of commercial power vendors as well as potential investors by helping them "determine how new capacity competes against existing capacity, and the response of the electric generators to the imposition of environmental controls on conventional pollutants or any limitations on greenhouse gas emissions."[64] Their assessments include the traditional power generating sources of coal, natural gas, nuclear, and hydroelectric power. But the reports also address alternative and renewable sources such as biomass, wind, geothermal, and solar (thermal and photovoltaic), which are becoming more common as they mature in technology and affordability.

Table 2 was developed using EIA estimates for terrestrial photovoltaic power plants, which have been scaled up to provide 1-GW capacity. However, Earth-bound photovoltaic plants do not have the constant sunlight exposure of an SBSP system. Thus, to make the values more comparable to the estimates in Table 1, an average capacity factor (the time a plant can actually produce a capacity) of 20 percent is used to increase the cost to the equivalent of full-time operation.[65] For comparison purposes, the estimates in both Tables 1 and 2 are set to a common reference year for dollar value (2013). Interestingly, the estimated cost range of $20.3-20.9 billion for a 1-GW terrestrial photovoltaic capacity is about the same as the $20 billion estimate for the first full-scale SPS-ALPHA satellite (which does not include the terrestrial and support systems of the SBSP system).

It may be difficult to convince investors to commit to an SBSP system that is undeveloped, unproven, and high risk when terrestrial systems that tap into the same renewable energy source already exist. Further, as the technology matures and applications spread, the capital costs for terrestrial photovoltaic decreased by 22 percent between 2010 and 2013. Environmentally, the systems are very clean, rated as generating zero amounts of nitrous oxide, carbon dioxide, or sulfur dioxide.

Nominal Plant Capacity	Overnight Capital Cost			Annual Operations & Maintenance Cost			30-year O&M Cost (1-GW Continuous)
	Each Plant	1-GW Capacity	1-GW Continuous	Each Plant	1-GW Capacity	1-GW Continuous	
20-MW (50 plants for 1-GW)	83.7 M	4.2 B	20.9 B	0.6 M	27.6 M	137.8 M	4.1 B
150-MW (7 plants for 1-GW)	581.0 M	4.1 B	20.3 B	3.7 M	25.9 M	129.6 M	3.9 B

Table 2. Terrestrial Solar Photovoltaic Power Plant Cost, 2013 Estimates.[66]

Regulatory Factors.

In addition to the cost of construction and operation, SBSP systems will most likely face considerable costs related to compliance with regulations of many forms. For simplicity, we can assume that the environmental regulations for construction of the SBSP terrestrial portion will be similar to that of a solar photovoltaic power generating plant. However, the routine and continuous propagation of high-energy-density microwave or laser beams require significant coordination. System radio frequencies will have to be evaluated and approved by the Federal Communications Commission (FCC) as well as the International Telecommunications Union (ITU). The potential of interference with civil aviation must be addressed by the Federal Aviation Administration (FAA) and the International Civil Aviation Organization (ICAO).[67] In its evaluation of NASA's SERT program, the National Research Council stated that "environmental, health, and safety issues are now recognized as essential concerns to be addressed as early in a program as possible."[68] This evaluation may include studies by the Environmental Protection Agency and the Occupational Safety and Health Administration that address the effects of long-term exposure to wireless energy transfer on plants, animals, and humans.

There will also be policy, legal, and regulatory considerations for the SBSP space system. The transportation of SBSP satellites to orbit will require coordination with the Department of Commerce and the Department of State for proper licensing and enduring compliance with the International Traffic in Arms Regulations (ITAR). Launch and early orbit operations should be coordinated with the proper authorities within DoD, such as the U.S. Strategic Command Joint Force Component Commander for Space. Also, investors must realize that the geosynchronous orbit belt around Earth should be considered a limited resource. GEO positions are allocated by the United

Nations ITU and are very competitive for services such as telecommunication and direct broadcasting. Figure 4 depicts the change in the GEO population in both volume and complexity from 1999 to 2011, during which period there were on average 26 spacecraft placed into GEO annually.[69]

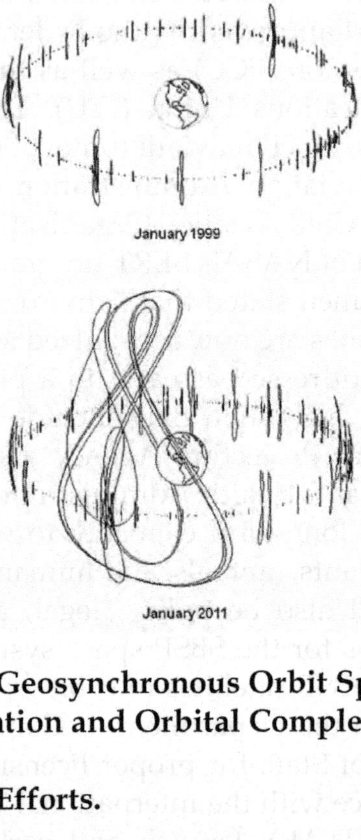

Figure 4. Geosynchronous Orbit Spacecraft Population and Orbital Complexity.[70]

International Efforts.

The pursuit of practical SBSP systems is not exclusive to the United States. In 2001, the National Research Council noted significant:

> international involvement in SSP [space solar power] and found an optimistic global picture. Japan, France,

> Canada, Russia, Ukraine, Georgia, Italy, Belgium, Germany, India, Netherlands, China, and Singapore are among the countries engaged in at least some facets of SSP research, development, and technological demonstration.[70a]

A decade later, the worldwide involvement remained strong, as demonstrated in the diversity of participants in the IAA Space Solar Power study, which included members from Japan, France, Norway, Canada, Germany, and the European Space Agency.

Japan has one of the most serious SBSP development programs, having formally incorporated a Space Solar Power Program into its Basic Plan for Space Policy in 2009. This plan included a 5-year development and utilization plan:

> [The Japanese] Government will examine the system for the development of space solar power program from a comprehensive point of view in collaboration with related institutions, and also conduct demonstration of technologies for the energy transmission technology in parallel. Based on the result, Government will conduct ample studies, then start technology demonstration projects in orbit utilizing 'Kibo' or small sized satellites within the next 3 years to confirm the influence in the atmosphere and system check.[71]

The Japan Aerospace Exploration Agency (JAXA) is spearheading an SBSP effort with "a technology road map that suggests a series of ground and orbital demonstrations leading to the development in the 2030s of a 1-gigawatt commercial system."[72] Some of its intermediate milestones include a 1-KW satellite experiment by 2017; a 2-MW satellite experiment by 2024; and a 200-MW demonstration power station by 2028.[73]

The ESA has conducted studies and development activities for several decades. Its approach has included objectives to identify "possible synergies between ground and space based power generation solutions" as well as other potential roles for SBSP systems, such as space exploration.[74] In 2003, the ESA published a program to evaluate solar power from space in three phases: general viability; system architecture level trade-off; and technology focus and demonstrator mission selection.[75]

However, the work of the ESA Advanced Concept Team during the validation phase determined that while the SBSP systems were technically feasible, "terrestrial plants [are] advantageous over space plants until [outputs of] several tens of GWe" and that "preliminary data show little to no advantage for European only based concepts."[76] Accordingly, while ESA currently has not progressed to a larger SBSP program, countries within the European Union have embraced terrestrially based solar photovoltaic plants. These efforts have proceeded better than planned; the result of "cumulative installed capability at the end of 2010 was over 29-GW, almost 10 times the original target."[77] Building on this success, the goal among 26 member states is to have this cumulative photovoltaic power generation up to 84.5-GW in 2020; this is equivalent to the output of approximately 100 Three Mile Island-class nuclear power plants.[78]

Other countries that have expressed interest in SBSP systems include Russia, China, and India. The China Academy of Space Technology is reported to have a five-step roadmap that culminates with a commercial SBSP satellite in GEO by 2050. There is a detailed proposal for a U.S.-India cooperative development, but it has received little serious attention.[79]

OPERATIONAL ASSESSMENT

This section explores the strategic considerations for SBSP systems within the general context of national space operations. It then examines potential garrison-level applications and compares these to the current plans of the Army's Energy Initiatives Task Forces to integrate terrestrial-based photovoltaic power into the energy systems of several major installations. Finally, this section briefly explores possible SBSP applications to support remote operating locations.

Strategic Considerations.

Is there a compelling national security requirement for SBSP systems? No such explicit mandate exists in U.S. capstone space strategy documents, nor is there any reasonable objective that implicitly points to such systems as having any priority in current plans. The *National Security Space Strategy* posits that "the current and future environment is driven by three trends—space is becoming increasingly **congested**, **contested**, and **competitive**."[80] The growth of Earth's orbital population depicted in Figure 5 clearly shows the increasing congestion and its consequences, such as over 4,500 pieces of orbital debris resulting from a Chinese anti-satellite (ASAT) test in 2007 and the collision of a Russian satellite with a U.S. commercial Iridium satellite in 2009. Even though the population has reduced somewhat since then due to the routine reentry of such debris, the July 2014 orbital population still stands at 16,900, of which only 3,812 objects are payloads.[81]

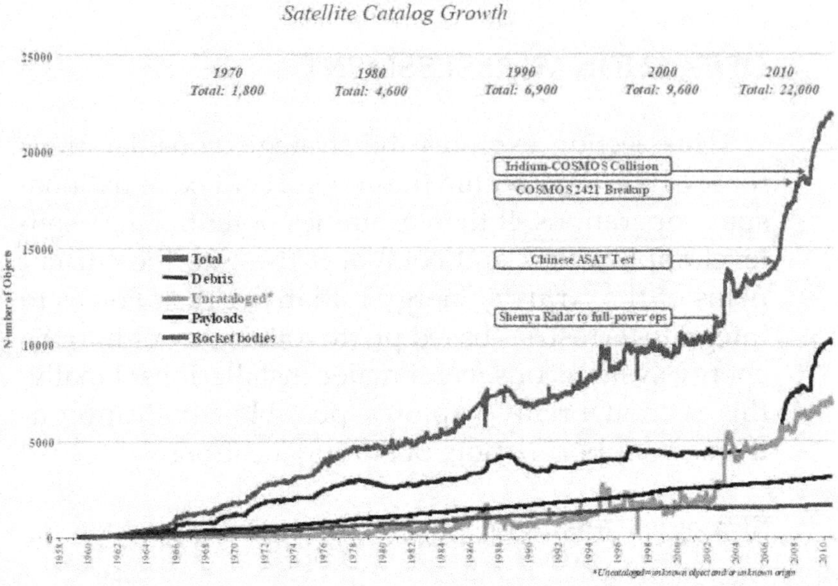

Figure 5. Earth Satellite Catalog Growth (1958-2011).[82]

Any additional satellites operating in this environment will only exacerbate the risk of safe operations for all other satellites. So perhaps the compelling question for proposed SBSP systems is this: Do they have a mission unique enough to justify their operation in this environment? Or, perhaps, stated another way, are the capabilities and benefits provided by SBSP systems exclusive to space or can they be accomplished on the ground, albeit perhaps more inefficiently? Certain space missions, such as surveillance, weather observation, and navigation spacecraft must leverage the ultimate high ground of space. Like these satellites, SBSP systems would have to operate in the severe and dynamic environment of space radiation, charged particles, and planetary debris. They would also be vulnerable to physical, electromagnetic, and cyber attacks. Finally, any contemplated use of SBSP systems would have to address the reasonable concern regarding the perceived ability of such systems to be weaponized.

But the competition for resources that SBSP systems must overcome is not restricted merely to gaining a location in orbit; these systems must also compete for supporting infrastructure with limited capacity, such as space lift, ground-tracking support, and satellite command facilities. Finally, the SBSP systems would have to compete for government budget resources in an era of desired fiscal austerity. For example, the $50 billion cost for SPS-ALPHA development and first 1-GW satellite (see Table 1) is more than the program cost of either a CVN 78 *Gerald R. Ford*-class aircraft carrier ($36.0 billion). It is also more than the United States Air Force's new KC-46 tanker aircraft ($44.5 billion) and significantly more than the $19 billion Space Based Infrared Satellite System (SBIRS High), which includes GEO satellites necessary for missile warning and battlefield awareness missions.[83]

It is also important to consider that the SBSP concept does not represent the only potential game-changing possibilities for energy security. Solving such problems as superconducting material or fusion energy generation would radically change the availability and use of electrical power on Earth. However, certain SBSP-related technologies, such as wireless energy transfer, merit further development; such systems could be used to distribute power to remote locations with limited infrastructure.

Garrison Applications.

Considering the expense and risk associated with current SBSP concepts, one may wonder if there are alternative options for using solar power as part of an overall energy portfolio. On September 15, 2011, Secretary of the Army John McHugh established the Energy Initiatives Task Force (EITF) "to serve as a one-stop shop for the development of cost-effective large-scale Army renewable energy projects."[84] The program was designed to address Army goals and federal mandates for enhanced energy security with an initial investment estimated at $7.1 billion over 10 years to generate 2.1 million-MW hours of power annually[85] (roughly the output of a 240-MW plant operating continuously). The current goal of the EITF is to deploy 1-GW of renewable energy projects by 2025 using solar, wind, biomass, and geothermal energy generation technologies.[86]

Facility	Facility Power Requirements (peak)	Planned Photovoltaic Power	Land Usage (acres)
Fort Detrick, MD[87]	40-MW	18.6-MW	80
Georgia 3 x 30[88] - Fort Benning - Fort Gordon - Fort Stewart	 73-MW 37-MW 62-MW	 30-MW 30-MW 30-MW	 250 250 250
Fort Huachuca, AZ[89]	23-MW	18-MW	155
Fort Irwin, CA[90]	28-MW	15-MW	600
Redstone Arsenal, AL[91]	75-MW	10-MW	188
Fort Stewart, GA (lease)[92]	62-MW	20-MW	111
TOTAL	**400-MW**	**171.6-MW** (42.9 percent peak)	**1884** (7.6 square kilometers)

Table 3. Army Sites for Terrestrial Solar Photovoltaic Power Plants.

Table 3 summarizes the current EITF efforts related to energy production using photovoltaic arrays at seven different Army garrisons in the continental United States. When fully implemented, the cumulative output of these plants will be about 172-MW, providing over 300,000-MW hours annually when operating at 20 percent average capacity.[93] The longer-range EITF goal is to have 505-MW of photovoltaic capability available to Army garrisons by 2020. Comparing the land required for the photovoltaic array to that required for the rectifying antenna of the 1979 NASA/DoE concept (see Figure 2), we find that the photovoltaic array can provide the equivalent of 1-GW continuous power using about 221-square KM compared to 102-square KM for the rectenna.[94] However, since the photovoltaic system does not require a 50-square-KM satellite in geosynchronous orbit, one could argue that if the 1979 NASA/DoE vision had been followed, it would now be eclipsed by the advanced state of the art of terrestrial photovoltaics.

Other factors also make terrestrial photovoltaic systems attractive. Their structure is co-located with the garrison that will use the power, thus reducing the extent of long-distance power distribution grid requirements. The distributed deployment also provides a more robust system, which is less susceptible to single-point failures or attacks. Also, since all of the system components are on the ground, the system can be more accessible for maintenance at lower cost as well as more flexible to incorporate design improvements or technology insertion opportunities.

Field Applications.

As discussed earlier, the NRL conducted a cursory review of possible military applications for SBSP systems; a summary table of its findings is included in the Appendix 1. While the report rates the use of SBSP systems as possible for forward operating base (FOB) applications, the NRL analysis states that the infrastructure and support personnel requirements do not offer significant advantages over traditional generator systems that use high energy-density fuel and are portable and well understood by operators. Assuming that typical FOB power requirements are no more than 5-MW, it is reasonable to question why the FOB would need the elaborate and expensive support from a GW-class SBSP system.[95] Even if a single GEO satellite could power multiple FOBs, it would only utilize a fraction of its potential capability.

Other studies have explored the possible use of mission orbits other than GEO for SBSP satellite applications. A 2009 Lawrence Livermore National Laboratory (LLNL) study explored "a space-based system capable of delivering 1-MW of energy to a terrestrial receiver station, via a single unmanned commercial launch into Low Earth Orbit (LEO)" for an estimated cost of $500 million.[96] But the operational effectiveness of such a system is severely limited by the LEO orbit, since the satellite only would be visible to the ground station for short periods. A 2012 thesis study at the Naval Postgraduate School calculated the actual viewing windows that a single ground power receiving station would have for SBSP systems stationed in four different orbits. Figure 6 depicts these orbits and provides their respective view times. The operational utility of the LLNL concept would be limited to less than 1 hour of power transmission each day, and this time would be distributed over five different orbital passes.[97] While using a highly elliptical orbit may provide up to 14.6 hours of satellite visibility each day, the use of any orbit below GEO would require significantly more sophisticated satellite attitude control and mission planning, since the orientation of the satellite relative to the ground station would be constantly changing at rates far above those in a GEO orbit.

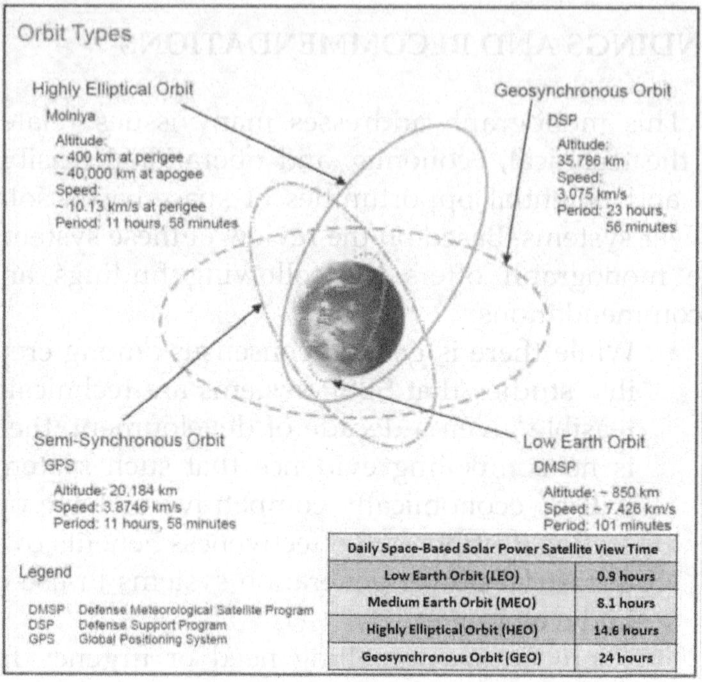

Daily Space-Based Solar Power Satellite View Time	
Low Earth Orbit (LEO)	0.9 hours
Medium Earth Orbit (MEO)	8.1 hours
Highly Elliptical Orbit (HEO)	14.6 hours
Geosynchronous Orbit (GEO)	24 hours

Figure 6. Potential Space-Based Solar Power Satellite Orbital Parameters.[98]

Potential niche use of SBSP systems to provide power for field applications, such as ultra-long duration unmanned vehicles or special operations forces, is certainly an attractive capability for warfighters.

But even if basic technical and economic challenges of SBSP systems are overcome, the remaining issues of power efficiency, beam-pointing controls, and beam-exposure hazards make such uses unlikely in the near future.[99] Even if wireless energy transfer is perfected from terrestrial sources, it use will require a serious review of implications in the areas of doctrine, organization, training, materiel, leadership and education, and personnel and facilities.

FINDINGS AND RECOMMENDATIONS

This monograph addresses many issues related to the technical, economic, and operational feasibility and potential opportunities of space-based solar power systems. Based on the review of these systems, the monograph offers the following findings and recommendations:

- While there is general consensus among credible studies that SBSP systems are technically feasible given a decade of development, there is no compelling evidence that such systems will be economically competitive or have operational utility and effectiveness benefits over terrestrial power generation systems in use or in development.
- There is no compelling need or urgency for SBSP systems articulated in current national security strategies or future roadmaps.
- Most reports that strongly advocate SBSP as a technically viable and economically competitive technology for energy have substantial biases for the system benefits and underplay many potential challenges and hazards. Also, the cost estimates of these reports focus on the

capital costs of the satellite portion of the overall system, which often make questionable assumptions regarding the unprecedented availability of space lift capability at significantly reduced cost. The credibility of such reports would be improved if they incorporated independent life-cycle cost estimates.

- SBSP should be considered part of a larger energy portfolio vice the exclusive energy approach advocated by some studies.
- The Army Energy Initiatives Task Force plan for terrestrial solar photovoltaic power is a prudent investment that will provide many of the benefits of SBSP systems in less time and with less risk.
- The technology of wireless energy transfer via microwave or laser beams has potential merit apart from SBSP application. Investing in the development of this technology may enable operations in remote and forward locations.

CONCLUDING REMARKS

Energy security is an issue that continues to become more acute as global populations grow and limited fossil fuel reserves shrink. The promises of space-based solar power for clean and unlimited energy for all humankind are certainly appealing. But the reality is that such systems always seem to be seen as just 10 years away by their advocates. While significant progress continues in the enabling technologies that will make SBSP systems economically viable and competitive power generators, there is no compelling evidence that such systems will provide the best energy solution. Considering the austerity of current federal budgets, the Army's evolutionary approach to incorporating a diverse portfolio of renewable energy sources in distributed locations seems more prudent than placing significant amounts of resources in high-risk ventures such as SBSP systems. Perhaps in a decade or so, there will be technological breakthroughs that will fully support practical SBSP systems. But it is also possible that within that decade there may be breakthroughs such as fusion energy exploitation, which will make SBSP systems obsolete before they are even fielded.

ENDNOTES

1. William Ledbetter, "An Energy Pioneer Looks Back: An Inspiring Conversation with Dr. Peter Glaser," *Ad Astra,* Vol. 20, No. 1, Spring 2008, pp. 26-27.

2. Peter Glaser, "Method and Apparatus for Converting Solar Radiation to Electrical Power," U.S. Patent No. 3,781,647, Alexandria, VA: U.S. Patent and Trademark Office, December 25, 1973.

3. *Ibid.,* p. 1.

4. David J. Chow, "Exploring the Feasibility of Providing Electrical Power to Remote Bases Via Space-Based Solar Power Satellites," Thesis for Master of Science in Systems Engineering Management, Monterey, CA: Naval Postgraduate School, June 2013, pp. 19-26. This section of the thesis includes a detailed systems engineering functional analysis of an SBSP system, broken down by functional, physical, and allocated architectures.

5. National Aeronautics and Space Administration, "Space-Based Solar Power Conversion and Delivery Systems Study, Vol. 1: Executive Summary" report NASA-CR-150294, Princeton, NJ: Econ Incorporated, March 31, 1977. The objectives of the study were stated as (p. 2):

With the above understanding in mind, the objectives of this study are to answer, sequentially, three questions:

1. Can it be done?
2. Should it be done?
3. How should it be done?

The first question addresses the basic technical and economic feasibility of the satellite solar power concept. Its purpose is to verify that the technology is at a point from which a development program may commence. The second question addresses the larger economic issue of whether a justification exists for undertaking the first phase of a satellite solar power development program. The last question addresses the level of effort and focus that results in an effective development program.

The full report had four additional volumes:

Vol. II, Engineering Analysis of Orbital Systems (NASA-CR-150295);

Vol. III, Analysis of the Microwave Power Transmission System Interfaces and Power Beam Ionospheric Effects (NASA-CR-150296);

Vol. IV, Analysis of Photovoltaic Energy Conversion Systems (NASA-CR-150297);

Vol. V, Economic Analysis (NASA-CR-150298).

6. "Three Mile Island Generating Station" fact sheet, Chicago, IL: Exelon Corporation, available from *www.exeloncorp.com/assets/energy/powerplants/docs/TMI/fact_tmi.pdf*, accessed September 12, 2014.

7. "Space-Based Solar Power Conversion and Delivery Systems Study, Vol. 1: Executive Summary" report, p. 2.

8. National Aeronautics and Space Administration, "Space-Based Solar Power Conversion and Delivery Systems Study, Vol. V: Economic Analysis" report NASA-CR-150298, Princeton, NJ: Econ Incorporated, March 31, 1977, p. 4.

9. *Ibid.*, pp. 8-10.

10. Department of Energy, "The Final Proceedings of the Solar Power Satellite Program Review," DoE/NASA Satellite Power System Concept Development and Evaluation Program, Washington, DC: DoE Office of Energy Research, July 1980.

11. *Ibid.*, p. 162.

12. *Ibid.*, p. 143.

13. John C. Mankins, "A Fresh Look at Space Solar Power: New Architectures, Concepts and Technologies," Paper IAF-97-R.2.03, Turin, Italy: 48th International Astronautical Congress, October, 1997, p. 1, available from *www.spacefuture.com/pr/archive/a_fresh_look_at_space_solar_power_new_architectures_concepts_and_technologies.shtml*, accessed on September 15, 2014.

14. *Ibid.*, p. 13.

15. National Research Council, *Laying the Foundation for Space Solar Power: An Assessment of NASA's Space Solar Power Investment Strategy*, Washington, DC: National Academy Press, 2001, p. 1.

16. *Ibid.*, p. 12-17. The report listed the NASA cost assumption of $800 per kilogram to reach GEO, a value that would require cost reductions greater than an order of magnitude. While this is not stated explicitly in the report, one can reasonably assume that the dollar values were based on its $2,000 value.

17. *Ibid.*, p. 8.

18. National Space Security Office, "Space-Based Solar Power as an Opportunity for Strategic Security: Phase 0 Architecture Feasibility Study," Interim Assessment Release 0.1, Washington, DC: National Space Security Office, October 10, 2007, p. i.

19. *Ibid.*, p. 5.

20. *Ibid.*, p. 9.

21. *Ibid.*, Appendix B-Demonstration Roadmap, pp. B-1 – B-5.

22. *Ibid.*, p. 31.

23. *Ibid.*, Appendix C-Business Case Analysis, pp. C-1 – C-2.

24. National Aeronautics and Space Administration "Advanced Space Transportation Program: Paving the Highway to

Space," fact sheet, Huntsville, AL: Marshall Space Flight Center, no date, available from *www.nasa.gov/centers/marshall/news/background/facts/astp.html*, accessed September 16, 2014.

25. National Space Security Office, "Space-Based Solar Power as an Opportunity for Strategic Security: Phase 0 Architecture Feasibility Study," p. 41-42.

26. W. Neil Johnson *et al.*, "Space-based Solar Power: Possible Defense Applications and Opportunities for NRL Contributions," Report NRL/FR/7650—09-10, 179, Washington, DC: Naval Research Laboratory, October 23, 2009, p. 1.

27. *Ibid.* The report's details regarding feasibility were:

> *1.1.1 Concept Feasibility*
>
> The NRL SBSP Study Group concurs with the conclusions of the numerous studies of preceding decades that the SBSP concept is technically feasible but that there remain significant system risks in many areas.
>
> The Group concurs that SBSP offers one of several possible solutions to the energy independence and dominance of our country and our military, and that those alternative solutions (including terrestrial solar, nuclear, and wind) must be an integral part of the solution.
>
> Safe power densities for wireless energy transmission generally restrict applications to large, relatively immobile, receiver sites.
>
> Capital, launch, and maintenance costs remain significant concerns in the economics of fielding a practical SBSP system, an analysis of which is beyond the scope of this monograph.

28. *Ibid.*, p. 2. For Relevant Research Areas, the report noted that:

> NRL has world class competencies in many of the engineering core areas necessary for SBSP risk reduction: thermal management, space structures, space robotic assembly, photovoltaics, RF (radio frequency) amplifier technology, energy storage and management, propulsion, and spacecraft and system engineering.

29. *Ibid.*, p. 3.

30. John C. Mankins and Nobuyuki Kaya, eds., *Space Solar Power, The First International Assessment of Space Solar Power: Opportunities, Issues, and Potential Pathways Forward*, Paris, France: International Academy of Astronautics, August 2011, p. 183.

31. *Ibid.*, p. xv.

32. *Ibid.*, p. 163.

33. *Ibid.*, p. 159.

34. John C. Mankins, "SPS-ALPHA: The First Practical Solar Power Satellite via Arbitrary Large Phased Array (A 2011-2012 NASA NIAC Phase 1 Project)," Santa Maria, CA: Artemis Innovation Management Solutions LLC., September 15, 2012, p. 17.

35. *Ibid.*, p. 47.

36. *Ibid.*, p. 106.

37. *Ibid.*, p. 105. The author (John C. Mankins) of this report went on to write the book, *The Case for Space Solar Power*, Houston, TX: The Virginia Edition, Inc., 2014. Chapter 7 of the book, "Low-Cost Space Transportation," provides an extensive and thorough review of the complex factors surrounding this challenge.

38. Mankins and Kaya, eds., *Space Solar Power, The First International Assessment of Space Solar Power: Opportunities, Issues, and Potential Pathways Forward*, p. 58. Table 4-1 of this report summarizes the SBSP key technology requirements from the IAA study in three categories: Class A-concept specific; Class B-generic space systems; and Class C-supporting infrastructures.

39. *Ibid.*, pp. 217-234 (Appendix E). This section of the report, "Technology Readiness and Risk Assessment," provides a detailed assessment of critical technological challenges from the IAA study. It also describes the methodologies of Research and Development Degree of Difficulty (R&D3) as well as "Technology Readiness Levels" (TRLs) that were originally developed by NASA and now are widely used in defense acquisition as well. The nine TRLs' progress observed (TRL-1) up to actual system

proven in successful operation (TRL-9). The goal of many technology development programs is to reach TRL-6, which is "a representative model or prototype system tested in a relevant environment" (p. 218).

40. Brian C. Busch, "Space-Based Solar Power System Architecture," Thesis for Master of Science in Space Systems Operations, Monterey, CA: Naval Postgraduate School, December 2012, p. 49. In Table 18 of his thesis, Busch compares the theoretical, laboratory, and production efficiencies for five common photovoltaic materials. He also provides the time needed for them to degrade in performance by 15 percent.

41. Johnson *et al.*, "Space-Based Solar Power: Possible Defense Applications and Opportunities for NRL Contributions," p. 15. Table 4 of this report also compares the efficiencies for 11 different types of solar cells (p. 23).

42. Chow, "Exploring the Feasibility of Providing Electrical Power to Remote Bases Via Space-Based Solar Power Satellites," pp. 32-34.

43. Kyra Wiens, "Solar Power When It's Raining: NRL Builds Space Satellite Module to Try," NRL News Release, Washington, DC: NRL, March 12, 2014, available from *www.nrl.navy.mil/media/news-releases/2014/solar-power-when-its-raining-nrl-builds-space-satellite-module-to-try*, accessed on September 9, 2014. Other areas of SBSP research that NRL can support are included in Table 12 (p. 77) of W. Neil Johnson *et al.*, "Space-based Solar Power: Possible Defense Applications and Opportunities for NRL Contributions."

44. Mankins, "SPS-ALPHA," p. 19.

45. Busch, "Space-Based Solar Power System Architecture," pp. 56-64. In this section of his thesis, Busch evaluates four existing commercial GEO satellite buses for potential SBSP application.

46. Chow, "Exploring the Feasibility of Providing Electrical Power to Remote Bases Via Space-Based Solar Power Satellites," p. 33.

47. National Research Council, *Laying the Foundation for Space Solar Power*, pp. 40-41.

48. Chow, "Exploring the Feasibility," p. 34.

49. National Research Council, *Laying the Foundation*, pp. 42-43.

50. Busch, "Space-Based Solar Power System Architecture," p. 74. Busch estimates that the construction of a GW-class satellite would require more than 120 heavy-lift space launches.

51. Department of Energy, "The Final Proceedings of the Solar Power Satellite Review," p. 104.

52. Sun Xin *et al.*, "Financial and Organizational Analysis for a Space Solar Power System: A Business Plan to Make Space Solar Power a Reality," A Multicultural Team Project for Master of Business Administration in Aerospace Management, Toulouse, France: Toulouse Business School, May 18, 2009, p. 60. The study assumes a market price for electricity of 0.08 Euros, which converts to slightly more than 10 cents (U.S.) An informal search of numerous electrical utility rate comparison websites in Pennsylvania showed rates to be within about 10 percent of this value.

53. Samuel H. Williamson, "Seven Ways to Compute the Relative Value of a U.S. Dollar Amount," available from *www.measuringworth.com/uscompare/relativevalue.php*, accessed on September 17, 2014. This website included a calculator that provides a range of relative values of the U.S. dollar between two selected years. The range is set by the lowest and highest values that result from seven different methods of comparison. For Table 1, the lowest value of each range was used to give the lowest cost for comparison. The calculated ranges used in calculating Table 1 values were: the 1977 to 2013 dollar range was 3.04 to 8.04; the 1997 to 2013 dollar range was 1.37 to 1.95; the 2007 to 2013 dollar range was 1.10 to 1.16; the 2009 to 2013 dollar range was 1.04 to 1.06; and the 2012 to 2013 dollar range was 1.00 to 1.04. The B in Table 1 represents billions.

54. Department of Energy, "The Final Proceedings of the Solar Power Satellite Review," p. 104; and National Aeronautics and Space Administration, "Space-Based Solar Power Conversion and Delivery Systems Study, Volume V: Economic Analysis," p. 4. The $89.0 billion for satellite development was derived by subtracting the total cost ($102.5 billion) from the embedded first unit cost ($13.5 billion).

55. Mankins, "A Fresh Look at Space Solar Power: New Architectures, Concepts and Technologies," pp. 8 and 11.

56. National Research Council, *Laying the Foundation for Space Solar Power*, p. 17.

57. National Space Security Office, "Space-Based Solar Power as an Opportunity for Strategic Security: Phase 0 Architecture Feasibility Study," p. B-1.

58. Mankins, "SPS-ALPHA: The First Practical Solar Power Satellite via Arbitrary Large Phased Array (A 2011-2012 NASA NIAC Phase 1 Project)," pp. 104-106.

59. Chow, "Exploring the Feasibility of Providing Electrical Power to Remote Bases Via Space-Based Solar Power Satellites," p. 47.

60. Xin *et al.*, "Financial and Organizational Analysis for a Space Solar Power System," p. 60.

61. Walmart financial information is from "Fortune 500 2014," available from fortune.com/fortune500/, accessed on September 20, 2014. NASA and DoE budgets are from "The President's Budget for Fiscal Year 2015," available from *www.whitehouse.gov/omb/budget/Overview*, accessed on September 20, 2014.

62. Xin *et al.*, "Financial and Organizational Analysis for a Space Solar Power System," pp. 60-65.

63. *Ibid.*, p. 108. The M and B in Table 2 represents millions and billions respectively and O&M represents operations and maintenance.

64. Energy Information Administration, *Updated Capital Cost Estimates for Utility Scale Electricity Generating Plants*, Washington, DC: Department of Energy, April 2013, p. 1.

65. John Miller, "What are the Capacity Factor Impacts on New Installed Renewable Power Generation Capacities?" available from *theenergycollective.com/jemillerep/450556/what-are-capacity-factor-impacts-new-installed-renewable-power-generation-capaciti*, accessed September 21, 2014. This blog article notes that "relatively new EIA [Energy Information Administration] data report shows that Wind and Solar PV [photovoltaic] capacity factors have been well below the projected state-of-art level" and recommended a value of 20 percent vice the EIA estimate of 25 percent. In keeping with the analysis methodology of this monograph, the author selected the 20-percent value in order to give the best case for SBSP systems.

66. *Ibid.*, p. 6. The relevant photovoltaic power plant estimates used from Table 1 of this source were:

20 MW plant:	$4,183 $/KW (capital)	$27.75 $/KW-yr (O&M)
150 MW plant:	$3,873 $/KW (capital)	$24.69 $/KW-yr (O&M)

Note that the term "overnight capital costs" refers to costs of the project if no interest is incurred during construction.

67. Mankins and Kaya, eds., *Space Solar Power, The First International Assessment of Space Solar Power: Opportunities, Issues, and Potential Pathways Forward*, pp. 72-73. Chap. 5, "SPS Policy and Other Considerations," addresses several regulatory and policy-related issues for the international community.

68. National Research Council, *Laying the Foundation for Space Solar Power*, p. 56.

69. Nicholas L. Johnson. "A New Look at the GEO and Near-GEO Regimes: Operations, Disposals, and Debris," IAC-11.A6.2.7 in *62nd International Astronautical Congress 2011 (IAC 2011) Proceedings*, Vol. 1 of 12, Cape Town, South Africa, October 3-7 2011, pp. 1932-1938.

70. *Ibid.*, p. 1938.

70a. National Research Council, *Laying the Foundation for Space Solar Power: An Assessment of NASA's Space Solar Power Investment Strategy*, pp. 25-26.

71. Nobuyuki Kaya, "New Basic Plans for Space Policy," *The Proceedings of the Panel Discussions on the Space Solar Power Systems*, Tsukuba, Japan: International Symposium on Space Technology and Science, July 10, 2009, pp. 33-42.

72. Susumi Sasaki, "How Japan Plans to Build an Orbital Solar Farm," IEEE Spectrum online, April 24, 2014, p. 1, available from *spectrum.ieee.org/green-tech/solar/how-japan-plans-to-build-an-orbital-solar-farm*, accessed on September 3, 2014.

73. *Ibid.*, p. 4. To accomplish its technology roadmap, JAXA established the following priorities:

> To complete and operate an electricity system based on such satellites, we would have to demonstrate mastery of six different disciplines: wireless power transmission, space transportation, construction of large structures in orbit, satellite attitude and orbit control, power generation, and power management. Of those six challenges, it is the wireless power transmission that remains the most daunting. So that is where JAXA has focused its research.

74. European Space Agency, "ESA Work on Solar Power from Space: Concluded and Ongoing Activities," Noordwijk, The Netherlands: ESA Advanced Concepts Team, January 2008, p. 2.

75. Leopold Summerer, "Solar Power from Space—European Strategy in the Light of Global Sustainable Development: Programme Plan," Noordwijk, The Netherlands: European Space Agency, July 8, 2003.

76. Leopold Summerer, "Solar Power from Space: Status and Perspectives," in *The Proceedings of the Panel Discussions on the Space Solar Power Systems*, Tsukuba, Japan: International Symposium on Space Technology and Science, July 10, 2009, p. 24.

77. Arnulf Jäger-Waldau, *Research, Solar Cell Production and Market Implementation of Photovoltaics*, Ispra, Italy: European Commission Joint Research Centre, October 2012, p. 36.

78. *Ibid.*, p. 38.

79. Peter Garretson, "Solar Power in Space?" *Strategic Studies Quarterly*, Vol. 6, No. 1, Spring 2012, pp. 97-123.

80. Robert M. Gates, Secretary of Defense, and James R. Clapper, Director of National Intelligence, *National Security Space Strategy: Unclassified Summary*, Washington, DC: U.S. Government Printing Office, January 2011, p. 1.

81. "Satellite Box Score as of 2 July 2014," *NASA Orbital Debris Quarterly News*, Vol. 18, No. 3, July 2014, p. 8.

82. Gates and Clapper, p. 1.

83. U.S. Government Accountability Office, "Defense Acquisitions: Assessments of Selected Weapon Programs," Report GAO-14-340SP, Washington, DC: U.S. Government Printing Office, March 31, 2014, pp. 18, 143.

84. Headquarters, Department of the Army, "Stand To! Energy Initiatives Task Force," Washington, DC: HQ, Department of the Army, September 20, 2011, p. 1, available from *www.army.mil/standto/archive/issue.php?issue=2011-09-20*, accessed on September 15, 2014.

85. *Ibid.*, p. 1.

86. John Lushetsky, "Renewables in the Federal Government: EITF Renewable Energy Overview," presentation at the 2013 World Energy Engineering Congress, Washington, DC, September 27, 2013.

87. "Fort Detrick, Maryland 18.6 MW Solar PV Project Overview," fact sheet, Arlington, VA: Army Energy Initiatives Task Force, November 2013, available from *www.asaie.army.mil/Public/ES/eitf/docs/FactSheet_FtDetrick.pdf*, accessed on September 12, 2014.

88. "Georgia 3x30: Bringing Large-scale Renewable Energy to Forts Stewart, Gordon, and Benning," fact sheet, Arlington, VA: Army Energy Initiatives Task Force, May 2014, available from *www.asaie.army.mil/Public/ES/eitf/docs/19May2014Ga3x30FactSheet.pdf*, accessed on September 12, 2014.

89. "Fort Huachuca Renewable Energy Project," fact sheet, Arlington, VA: Army Energy Initiatives Task Force, March 2014, available from: *www.asaie.army.mil/Public/ES/eitf/docs/FactSheet_FtHuachuca.pdf*, accessed on September 12, 2014.

90. "Fort Irwin, California 15 MW Solar PV Project Overview," fact sheet, Arlington, VA: Army Energy Initiatives Task Force, June 22, 2013, available Fort Irwin, CA, 15 MW Solar PV Project Overview," fact sheet, Arlington, VA: Army Energy Initiatives Task Force, June 22, 2013, available from *www.asaie.army.mil/Public/ES/eitf/docs/SolarPV-FortIrwinCalifornia.pdf*, accessed on September 12, 2014.

91. "Redstone Arsenal Solar Project Task Order Under MATOC," fact sheet, Arlington, VA: Army Energy Initiatives Task Force, July 2014, available from *www.asaie.army.mil/Public/ES/eitf/docs/RedstoneArsenalFactSheet.pdf*, accessed on September 12, 2014.

92. "Fort Stewart, Georgia Solar PV Lease," fact sheet, Arlington, VA: Army Energy Initiatives Task Force, January 2014, available from *www.asaie.army.mil/Public/ES/eitf/docs/FactSheet_FtStewart_01-2014.pdf*, accessed on September 12, 2014.

93. Using an average capacity value of 20 percent for Table 3 values is consistent with the assumptions used to generate Table 2 values.

94. Calculations for these numbers used an average capacity value of 20 percent for terrestrial photovoltaic (PV) systems. Thus, a 5-GW land PV system is needed to provide the same output as a 1-GW SBSP system operating continuously. Applying the direct ratio scaling from 171.6-MW to 5-GW to 7.6 square KM yields 221 square KM for the terrestrial system. The area for the rectenna depicted in Figure 2 is simply multiplying the semi-major (6.5-km) and semi-minor (5.0-km) axes values together with pi.

95. Xin *et al.*, "Financial and Organizational Analysis for a Space Solar Power System," p. 70. This report bases its FOB usage estimates on the details of FOB operations explored by the Air Force Research Laboratory. The report also describes a 1-MW Mobile Space-to-Ground Solar Power Station (MSGSPS) system that is expeditionary in design, using C-130 airlift transportation and standard cargo trucks.

96. A. M. Rubenchik *et al.*, "Solar Power Beaming: From Space to Earth," report LLNL-TR-412782, Livermore, CA: Lawrence Livermore National Laboratory, April 2009, pp. 1, 11.

97. Busch, p. 22.

98. *Joint Publication 3-14, Space Operations*, Washington, DC: Joint Chiefs of Staff, May 29, 2013, p. G-4. The Orbit Types graphic is from Figure G-1. Also see Busch, pp. 19-44. Chap. 3 of this thesis provides the detailed calculations for the satellite view times as shown in Figure 6.

99. Johnson *et al.*, "Space-based Solar Power," pp. 5-6.

APPENDIX 1

ASSESSMENT OF SPACE-BASED SOLAR POWER APPLICATIONS FOR MILITARY OPERATIONS SCENARIOS*

Military Operation Scenario	Rationale for SBSP	Feasibility		Notes	Earliest operational capability	Rough magnitude cost
		Technical	Economic			
Forward Operating Base Power	Reduce fuel convoys	Possible	Possible	Probably best SBSP defense app	>5 years	$10B+
Provide power to a ship or other large seaborne platform	Refuel from space	Possible	Possible	Almost certainly requires lasers and high power densities	> 5 years	$10B+
Bistatic radar il-lumination	Improve imaging	Possible	Possible	Feasible but expensive	> 5 years	$10B+
Provide power to a remote location for synthfuel produc-tion	Reduce infrastruc-ture	Possible	Possible	Requires transportation architecture that consumes synthfuel	> 5 years	$10B+
Power to Individual End Users	Reduce battery mass	Unlikely	Unlikely	Power inefficient, severe beam control, and safety challenges	> 10 years	?
Power for Dis-tributed Sensor Networks	Cover large area	Possible	Possible	Power inefficient	> 5 years	$10B+
Spaced solar power to non-terrestrial targets						
Satellite to satellite power transmission	Fractionate spacecraft	Possible	Possible	Significant techni-cal issues, ques-tionable utility	> 2 years	$50M+
Space to unmanned aerial vehicle (UAV) for dwell extension	Prolong dwell time	Possible	Possible*	*if used in con-junction with FOB power	> 5 years	$10B+
Terrestrial Wireless Power Beaming Applications Apart from SBSP						
Ship to shore power beaming	Increase flexibility	Possible	Possible	Attractive defense app, requires more study	> 1 years	$10M+
Ground to UAV for dwell extension	Prolong dwell times	Demonstrated	Possible	May be unneces-sary in light of recent UAV tech advances	> 1 years	$10M+

*Source: W. Neil Johnson *et al.*, "Space-based Solar Power: Possible Defense Applications and Opportunities for NRL Contributions," Report NRL/FR/7650—09-10, 179, Washington, DC: Naval Research Laboratory, October 23, 2009, p. 3 (Table 1, Investigation Summary).

U.S. ARMY WAR COLLEGE

Major General William E. Rapp
Commandant

*****

STRATEGIC STUDIES INSTITUTE
and
U.S. ARMY WAR COLLEGE PRESS

Director
Professor Douglas C. Lovelace, Jr.

Director of Research
Dr. Steven K. Metz

Author
Mr. Jeffrey L. Caton

Editor for Production
Dr. James G. Pierce

Publications Assistant
Ms. Rita A. Rummel

*****

Composition
Mrs. Jennifer E. Nevil

Made in the USA
Monee, IL
04 November 2026

67380391R00039